1st Symposium on Incoherent Light Sources

Loughborough, United Kingdom
8-11 April 1975

ISBN: 978-1-5108-4106-2

Printed from e-media with permission by:

Curran Associates, Inc.
57 Morehouse Lane
Red Hook, NY 12571

Some format issues inherent in the e-media version may also appear in this print version.

Printed by Curran Associates, Inc. (2017)

Foundation for the Advancement of the Science & Technology of Light Sources
FAST-LS
Belmayne House
99 Clarkehouse Road
Sheffield, United Kingdom
S10 2LN

www.fast-ls.org

Additional copies of this publication are available from:

Curran Associates, Inc.
57 Morehouse Lane
Red Hook, NY 12571 USA
Phone: 845-758-0400
Fax: 845-758-2633
Email: curran@proceedings.com
Web: www.proceedings.com

SYMPOSIUM

ON

INCOHERENT LIGHT SOURCES

Organized jointly by The Institute of Physics and Loughborough University of Technology in collaboration with The Institution of Electrical Engineers

LOUGHBOROUGH UNIVERSITY OF TECHNOLOGY

8 - 11 APRIL 1975

NOT FOR PUBLICATION

O U T L I N E S   O F   P A P E R S

The outlines of papers contained in this handbook are compiled for the convenience of those attending the conference and others interested in the same field of work. They are not for publication; they should not be quoted in the literature nor may they be reproduced in abstracting journals or similar publications since they do not necessarily relate to papers intended for publication.

Data contained in the outlines may only be quoted with the permission of the authors.

P U B L I C A T I O N   O F   P A P E R S

Contributors to the conference are invited to submit suitable original papers for publication in The Institute of Physics Journal of Physics Series.

For further information please write to the Managing Editor, Publishing Division, The Institute of Physics, 23 Marsh Street, Bristol BS1 4BT.

ORGANISING COMMITTEE

Professor J Raffle	-	University of Technology, Loughborough
Dr M A Cayless	-	Thorn Lighting Ltd, Leicester
Dr J R Coaton	-	Thorn Lighting Ltd, Leicester
Mr N J Phillips	-	University of Technology, Loughborough
Mr C E Williams	-	GEC Hirst Research Centre, Wembley, Middx.

GENERAL INFORMATION

VENUE

The Conference will be held in the Physics Department of the University of Technology, Loughborough. Resident participants will be accommodated in the Faraday Hall of Residence. A map of the University site is enclosed.

TRANSPORT

Rail

Participants travelling from London travel from St Pancras to Loughborough and then by taxi to the University. If overseas visitors travel to the East Midlands Airport, transport can be arranged to the University if they advise Mr Phillips, in the Physics Department, Loughborough, accordingly.

Road

Participants travelling to Loughborough by road should leave the M1 motorway at junction 23.

PARKING

Car parking facilities will be available at the University.

SYMPOSIUM OFFICE

A reception desk will be situated in the Faraday Hall from 16.00 hrs. on Monday 7 April 1975. From 09.00 hrs on Tuesday 8 April and for the remainder of the symposium the desk will be situated in the foyer of the Physics Dept.

MEALS

Meals will be served in Faraday Hall at the following times:

Breakfast	08.00 - 08.45
Dinner	18.30

Luncheon will be served in Faraday Hall at the times stated in the programme. Coffee and tea will be served in the Physics Department at the times shown in the programme.

RECEPTION

A sherry reception will be held in the common room of Faraday Hall at 20.00 hrs on the evening of Monday 7 April at which all participants will be welcome.

SYMPOSIUM DINNER

An informal dinner will be held in Faraday Hall at 19.30 hrs for 20.00 hrs on Thursday 10 April 1975. Tickets are enclosed for resident participants and non-resident participants who have ordered them.

VISITS

Visits will be arranged on the afternoon of Wednesday 9 April to:

(i) Thorn Lighting Ltd, Research and Engineering Laboratories

(ii) The Physics Laboratory, Loughborough University of Technology

(iii) Belvoir Castle

Coaches will leave from outside the Administration Buildings at 14.00 hrs for visits (i) and (ii).

MEDICAL CENTRE

There is a medical centre at the University and normally consultations can be made after 10.00 hrs. Emergency services can be provided if required outside normal surgery hours. Contact Mr N Phillips in Rutherford Lodge (extension 300).

VISUAL AIDS

A slide projector for 2" x 2" (5 cm x 5 cm) slides will be available together with an overhead projector and a 16 mm film projector. Slides should be handed to the projectionist at the beginning of the session in which they are to be shown. The organisers cannot accept responsibility for slides not collected.

SYMPOSIUM BADGE

A symposium badge is enclosed. Participants are asked to wear these for the duration of the symposium.

CANCELLATIONS

Participants are reminded that no payment can be refunded to those cancelling in whole or in part if such cancellations are received after 27 March 1975.

TELEPHONE MESSAGES

Urgent telephone messages for participants will be taken on:

050-93 63171 ext 5245 (physics department)
050-93 4907 (Faraday hall)

ENQUIRIES

Enquiries regarding the programme should be addressed to:

Mr N J Phillips
Dept of Physics
University of Technology
Loughborough, Leics. LE11 3TU

All other enquiries to:

The Meetings Officer
The Institute of Physics
47 Belgrave Square
London SW1X 8QX

PROVISIONAL PROGRAMME

TUESDAY 8 APRIL 1975

INCANDESCENT SOURCES

09.15 Introduction and Welcome

09.25 Invited Paper : DR B KOPELMAN (GTE Sylvania Inc, USA)

10.05 1 THERMOCHEMISTRY AND KINETICS OF THE TUNGSTEN BROMINE SYSTEM ...7
V W Goddard (Thorn Lighting Ltd, Leicester)

10.25 2 TRANSPORT IN THE TUNGSTEN FLUORINE SYSTEM 8
G Dittmer, A Klopfer and J Schroder (Philips, Aachen, Germany)

10.45 MORNING COFFEE

11.10 3 FLUORINE/FLUORIDE RESISTANT COATINGS OF FUSED QUARTZ....9
J M Rees (Thorn Lighting Ltd, Leicester), R Rothon (ICI Ltd, Runcorn, Cheshire)

11.30 4 BURNOUT MECHANISMS IN INCANDESCENT LAMPS....10
W Lems, H Kinkartz, W Lechner (Philips, Aachen)

11.50 5 INVESTIGATION OF TRANSPORT PROCESSES IN HALOGEN LAMPS BY MEANS OF RADIOACTIVE TRACER METHODS....11
L Bartha, T Geszti, L Kozma (Research Institute for Technical Physics of the Hungarian Academy of Sciences, Budapest)

12.10 6 ON CARBON AS A GETTER OF OXYGEN IN HALOGEN LAMPS....12
T Vicsek (Research Institute for Technical Physics of the Hungarian Academy of Sciences, Budapest)

13.00 LUNCH

DIAGNOSTICS AND SPECTROSCOPY

14.00 7 HEAT AND MASS TRANSPORT IN GAS-FILLED INCANDESCENT LAMPS....13
J Fitzgerald and E Fischer (Philips, Aachen)

14.20 8 PROJECTION HOLOGRAPHIC INTERFEROMETRY AS A TECHNIQUE FOR THE DIAGNOSTICS OF GAS-FILLED LAMPS....15
N J Phillips (University of Technology, Loughborough)

14.40 9 GAS-PHASE SEPARATION MECHANISMS IN LINEAR TUNGSTEN-HALOGEN LAMPS....16
J R Coaton (Thorn Lighting Ltd, Leicester)

15.00 10 SPECKLE PATTERN DIAGNOSTICS OF GASEOUS INTERIORS....N/A
J N Butters (University of Technology, Loughborough)

15.20 TEA

15.40 11 THREE DIMENSIONAL TIME-RESOLVED MEASUREMENTS....N/A
R F Stevens (NPL, Teddington)

16.00 12 TEMPERATURE AND VELOCITY FIELD BETWEEN CO-AXIAL HORIZONTAL CYLINDERS IN FREE CONVECTION....17
A Szokefalvi-Nagy, G Tichy, I Gaal (Research Institute of the Hungarian Academy of Science, Budapest)

16.20 13 SOME PROBLEMS OF FORMING THE IMAGE OF GASEOUS (PHASE) OBJECTS....18
J H Toyer, N J Phillips (University of Technology, Loughborough)

WEDNESDAY 9 APRIL

DIAGNOSTICS AND SPECTROSCOPY

09.00 Invited Paper : DR D D BURGESS (Physics Dept, Imperial College, London)

09.40 14 SPECTROSCOPIC TECHNIQUES FOR DIAGNOSTICS OF METAL HALIDE ARC LAMPS....19
J F Waymouth, W M Keeffe, W I Bamberg, P J Gardner, W Walter, D Work
(GTE Sylvania Inc, USA)

10.00 15 EMISSION LINES IN THE SPECTRA OF HALOGEN INCANDESCENT LAMPS....20
W J van den Hoet, E G Berns (Philips, Eindhoven)

10.20 16 SYCHROTRON RADIATION SOURCES....21
I H Munro (Daresbury Laboratory, UKAEA Warrington)

10.40 MORNING COFFEE

11.00 17 AUGER ELECTRON SPECTROSCOPY. AN ELECTRON SPECTROSCOPY FOR CHEMICAL ANALYSIS....22
Martha Thomas (GTE Sylvania Inc, USA)

11.20 18 AN APPROXIMATE IMAGE EQUATION FOR A NEW METHOD OF THE OPTICAL DIANGOSTICS OF A GLOW DISCHARGE COLUMN....23
A Sprachta (Tesla V.U.S.T. Prague)

11.40 19 PHOTON PRODUCTION AND INTENSITY CORRELATION....24
A A Walma (University of Sheffield)

12.00 20 A CONSISTENT ANALYSIS OF INCOHERENT RADIATION TRANSFER PROCESSES IN DISCHARGE LIGHT SOURCES....25
M A Cayless (Thorn Lighting, Leicester)

12.20 Close of morning session

12.30 LUNCH

AFTERNOON VISITS TO: Thorn Lighting Ltd, Research & Engineering Laboratories, Leicester
Physics Department, University of Loughborough
Belvoir Castle

THURSDAY 10 APRIL

DISCHARGE SOURCES

09.00 Invited Paper : D M KOEDAM (Philips, Eindhoven)

09.40 21 HALIDE COMPLEXES IN HIGH PRESSURE DISCHARGE LAMPS....26
A G Chalmers, D O Wharmby, F L Whittaker (Thorn Lighting Ltd, Leicester

10.00 22 ENHANCEMENT OF THE PARTIAL PRESSURE OF LOW VOLATILITY HALIDES IN HIGH PRESSURE DISCHARGE LAMPS....27
R Lorenz (Philips, Eindhoven)

10.20 23 EFFECTS OF MASS TRANSPORT ON ENERGY BALANCE IN METAL HALIDE ARCS....28
G J Kasek (GEC Cleveland, Ohio, USA)

10.40 MORNING COFFEE

11.00 24 AN ELEMENTARY ARC MODEL OF THE HIGH PRESSURE SODIUM LAMP....29
J F Waymouth (GTE Sylvania Inc, USA)

11.20 25 MEASUREMENTS OF ELECTRODE POWER DISSIPATION IN HIGH PRESSURE SODIUM LAMPS....30
P L Denbigh, D O Wharmby (Thorn Lighting Ltd, Leicester)

11.40 26 THE TIN HALIDE ARC....31
P C Drop, J J de Groot, A G Jack (Philips, Eindhoven)

12.00 27 C.W. AND PULSED ZINC AND CADMIUM VAPOUR ARC LAMP FOR UNDERWATER USE....32
M Geller (US Naval Electronic Laboratory Centre, San Diego, USA)

12.20 28 A NEW COMPACT FLUORESCENT TUBE....34
C E Williams (GEC Hirst Research Centre, Middlesex)

12.40 Close of morning session

13.00 LUNCH

ELECTRO-LUMINESCENCE

14.00 Invited Paper : PROFESSOR G F GARLICK (Dept of Physics, University of Hull)

14.40 29 MULTI-CHIP SEMI-CONDUCTOR LAMPS....35
R Hall, J R Fitzpatrick, B F Jones (Thorn Lighting Ltd, Leicester)

15.00 30 DEEP LEVEL LIMITED EFFICIENCIES IN GALLIUM PHOSPHIDE L.E.D'S....36
B Hamilton, T J Hayes, A R Peaker, B L Smith (Ferranti Ltd, Oldham)

15.20 TEA

15.40 31 TRANSIENT BEHAVIOUR OF LIGHT EMITTING DIODES....37
A Descombes (Federal Institute of Technology, Zurich)

16.00 32 FACTORS AFFECTING THE EFFICIENCY OF POWDER ZnS DCEL DISPLAYS....38
P H T Beatty (GEC Hirst Research Centre, Wembley)

16.20 Close of afternoon session

FRIDAY 11 APRIL

MATERIALS

09.00 Invited Paper : DR H RAWSON (University of Sheffield)

09.40 33 THE MICRO-STRUCTURAL AND MECHANICAL BEHAVIOUR OF TUNGSTEN WIRES AT ELEVATED TEMPERATURES....40
J H Day (Thorn Lighting Ltd, Leicester)

10.00 34 CARBON AS AN ALTERNATIVE FILAMENT MATERIAL 41
W J van den Hoek (Philips, Eindhoven)

10.20 35 THE EFFECT OF POROSITY AND GRAIN STRUCTURE ON THE LIFETIME OF INCANDESCENT LAMP FILAMENTS 42
O Horacsek, K Horacsek, L Bartha (Research Institute for Technical Physics, Hungarian Academy of Science, Budapest)

10.40 MORNING COFFEE

11.00 36 CLEANING OF TUNGSTEN FILAMENTS BY ION BOMBARDMENT 43
J M Walls (University of Loughborough)

11.20 37 MAGNESIUM ALUMINATE SPINELS AS LAMP ENVELOPE MATERIALS 44
P Hing (Thorn Lighting Ltd, Leicester)

11.40 38 INFRA RED REFLECTING FILTERS AND THEIR APPLICATION IN LIGHT PRODUCTI(....4
H Kostlin (Philips, Aachen)

12.00 39 INTERMEDIATE SEALS FOR LAMPS WITH FUSED QUARTZ ENVELOPES...N/A
T J Barry, L A Lay (National Physical Laboratory, Teddington)
D T Evans (Thorn Lighting Ltd, Leicester)

12.20 40 HIGH TEMPERATURE RESISTANCE OF INCANDESCENT FILAMENTS....N/A
I Gaal, J M Neugebauer, L Uray (Research Institute for Technical Physics of the Hungarian Academy of Sciences, Budapest)

12.40 Close of symposium

13.00 LUNCH

Paper 1

THERMOCHEMISTRY AND KINETICS OF THE TUNGSTON BROMINE SYSTEM

V W Goddard (Lamp Research Department, Thorn Lighting Ltd., Leicester)

Almost all the recently published work on tungsten-halogen systems has been biased towards pure thermodynamic considerations with a complete disregard for the kinetic considerations and the known chemistry of the tungsten halides.

The so called "Mass Balance" treatment is based on a few measured enthalpies, from which a considerable number of other data are derived by approximation methods, These methods give results which are at variance to known chemical properties, for instance, thermodynamical treatment shows that WO_2Br_2 is an important high temperature gas phase species in the tungsten bromine lamp; whereas from a thermochemical consideration it dissociates at $> 300^oC$.

It has been shown that the presence of oxygen is not necessary for the tungsten-bromine cycle to function efficiently but it is difficult to exclude this element therefore its effect on the cycle must be considered. Calorimetry and sealed tube experiments have shown that $WOBr_4$ is the important oxybromide species. Tungsten tetrabromide was originally thought to be an important gas phase bromide but this has been shown by calorimetry to undergo solid state decomposition at temperatures in excess of 280°C to form the dibromide and pentabromide. Tungsten dibromide has been shown to dissociate to the elements at temperatures around 800°C and there is evidence to suggest that it is a gaseous decomposition. The important gas phase tungsten bromide is the pentabromide and filament decomposition experiments have substantiated the measured thermodynamic parameters with respect to decomposition temperature.

The reactivity of tungsten pentabromode towards fused quartz has been studied in detail. Reaction occurs at temperatures $> 700^oC$ initially forming $WOBr_3$, which by a complex decomposition route, results in $WOBr_4$ being the major product. The rate of attack is dependent on the hydroxyl content of the fused quartz.

The kinetics of the reaction between tungsten and bromine have been studied. The results have been interpreted as showing that the major surface produced species changes from WBr_5 to WBr_2, with increase in temperature. The magnitude of the activation energy suggests that the bromides are formed by a stepwise addition of atomic bromine.

Paper 2

TRANSPORT IN THE TUNGSTEN-FLUORINE SYSTEM

G Dittmer, A Klopfer and J Schroder (Philips, Aachen, Germany)

The tungsten-fluorine system has been investigated with the aim of realizing a regenerative cycle for the filament material of incandescent lamps. The tungsten-fluorine cycle has been demonstrated to exhibit a preferential transport of tungsten from the colder to the hottest parts of the filament, a process which yields a homogeneous filament temperature but causes chemical attack of the cold leads.

In order to get an insight into the relevant processes occuring in a tungsten-fluorine lamp, the tungsten transport with fluorine and fluorine containing compounds was investigated. The experiments were performed on filaments under both steady-state and dynamic conditions in bulbs with chemically active atmospheres and in an inert gas stream containing fluorine respectively. From these experiments thermodynamic data of the lower valent tungsten fluorides were derived and the tungsten solubility in the gas phase was calculated as a function of temperature and pressure.

The tungsten solubility shows a maximum at about 2000 K where WF_4 is the predominant compound and drops towards higher temperatures to a minimum of about 0.1 of the maximum value. This decrease in solubility with increase in temperature leads to the transport from colder to hotter parts at temperatures above the maximum.

In order to obtain a true regeneration of the filament material, there has to be no net ransport of tungsten. Hence, attempts were made to fill up the solubility gap by adding compounds which release fluorine in this temperature range and thus raise the partial pressures of the lower valent tungsten fluorides. For this purpose volatile compounds such as SiF_4 and PF_3 were introduced into the system.

Another possibility is to add oxygen to the $W-F_2$ system which shifts the decrease in tungsten solubility to higher temperatures.

Paper 3

FLUORINE/FLUORIDE RESISTANT COATINGS FOR FUSED QUARTZ

J M Rees (Thorn Lighting Ltd, Melton Road, Leicester)

R Rothon (Imperial Chemical Industries Ltd, The Heath, Runcorn, Cheshire)

A major stumbling block to the production of a fluorine lamp is the reaction of the halogen with the fused quartz bulb wall. This results in the rapid deposition of tungsten onto the bulb with a consequent loss of light. Most conventional fluorine resistant glasses such as those based on phosphates or beryllium fluoride are sufficiently refractory for lamp making purposes.

A method is described by which fused quartz can be protected against fluorine attack by coating with aluminium/titanium phosphate based glasses which have melting points in excess of 1200°C. These glasses are applied from methanolic solutions of metal phosphate complexes, which, after drying, are cured at $\sim$ 400°C, in vacuo, to form the aluminium/titanium phosphate glass. Small apertures in the envelope, such as the lamp exhaust tube, $\sim$ 2mm diameter, can be sealed without destroying the continutity of the coating.

Results of fluorine containing lamps in which the bulb wall has been operating in excess of 500°C with a total gas pressure of 10 atomospheres argon show the feasibility of an improved lamp. The variation of tungsten deposition temperature with fluorine concentration shows good agreement with published kinetic and thermodynamic results. Lamp life is at present limited to $\sim$ 50h because of tail erosion.

Paper 4

BURN-OUT MECHANISMS IN INCANDESCENT LAMPS

W Lems, H Kinkartz, and W Lechner
(Philips Forschungslaboratorium Aachen GmbH, Aachen, West Germany)

Observations of the facetted structure of tungsten coils which have burned in gas-filled lamps led to the question whether these facetted single crystal regions can cause local variation of the energy balance and hence hot spots.

A series of tungsten coils (36 µm wire diameter, 130 turns, 1 atm. inert gas fillings) were burned with a.c. and d.c. at various temperatures. The temperature development along the coil was continuously registered and shortly before burn-out the lamps were switched off. The coils were subsequently investigated with a scanning electron microscope.

In the temperature range below 3000 K the coils were observed to possess strongly developed facets which were predominantly of the {100} type. When operated with d.c. these facets were somewhat more pronounced, and also showed the typical small "d.c. etch" facets. The inside of these coils was found to be lozenge-shaped, whereas the inside of the a.c. burned coil remained much rounder. The fact that the facets on the inside and outside are not parallel to one another leads to notches which are life determining for d.c. operated coils below 2700 K. An activation energy of about 60 kcal/mol was determined from the life tests.

Above 2700 K no difference in life is observed between a.c. and d.c. operated coils. As the facets are more strongly developed on the d.c. specimen , facets do not appear to influence life in this temperature range. In addition no correlation between temperature distribution and the facetted single crystal region was observed. The activation energy of the life controlling process is observed to correspond to that of evaporation: about 200 kcal/mol.

At temperatures above 3000 K, the facets are only weakly developed. Preliminary results indicate an activation energy which is appreciably higher than that for evaporation. Could diffusion of the dope substance potassium in tungsten play a role

The influence which inert gas pressure and wire diameter could have on the mechanisms, and thus on the temperature where they play a role, will be briefly discussed.

Paper 5

INVESTIGATION OF TRANSPORT PROCESSES IN HALOGEN LAMPS BY MEANS OF RADIOACTIVE TRACER METHODS

L Bartha, T Geszti, L Kozma
(Research Institute for Technical Physics of the Hungarian Academy of Science, Budapest, Hungary)

Not only the transport of tungsten atoms by the halogene cycle, but the transport of several other atoms, being present as construction materials or impurities, may take place in halogene lamps. Such processes were investigated by setting radioactive isotopes of different metals into halogen lamps. The kinetics of the transport processes and the steady state distribution of the isotopes have been measured.

It has been shown, that in tubular lamps /220 V, 1000 W, 50 hours designed lifetime/ only a relatively small fraction of the Fe and Co atoms will condense on the filament, but W and Ta condense almost quantitatively. Autoradiograms have proved that after condensation the latter atoms penetrate into the tungsten filament by lattice diffusion mechanism.

The distribution of the atoms along the filament transported from a point-like source has been also experimentally determined and the results were theoretically treated in a diffusional approximation.

Paper 6

ON CARBON AS A GETTER OF OXYGEN IN ALOGEN LAMPS

T Geszti and T Vicsek
(Research Institute for Technical Physics of the Hungarian Academy of Sciences, Budapest, Hungary)

The theory of quasi-steady state radial and axial transport in a halogen lamp is revised. Assuming a common diffusion coefficient for all gaseous components, the total amount of each atomic species X can be regarded as a single transported entity, the overall partial pressure P_X of which satisfies a diffusional-convectional transport equation. The redistribution of P_X among various compounds is determined by the mass action law equations, under the assumption that chemical reactions are fast with respect to transport.

The case of no net transport of an atomic species is described by a Daltonian spatially constant solution of the corresponding transport equation, supplemented by the boundary conditions of either the partial pressures being too low for condensation, or satisfying local equilibrium with already condensed reactive phases. The latter is obeyed even if transport is going on but physical and chemical evaporation is very fast.

For tungsten the resulting treatment is equivalent to the usual thermodynamical description. A new feature appears for carbon which does not satisfy a non-condensation boundary condition if usual amounts of it are present in the lamp. Therefore one has to admit the presence of carbon condensed on the wall, which must be in local equilibrium with carbon in the gas space. This places a serious limitation on the activity of carbon as a getter of oxygen. In particular, under usual conditions the constancies of P_C as well as P_O imply that close to the filament there is inevitably almost as much free oxygen present as there is CO_2 close to the wall. Therefore good gettering by means of carbon relies upon eliminating the formation of CO_2 as far as possible, which requirement is shown to demand first of all a wall temperature nowhere lower than ~ 800 °K. A new explanation is offered to a related experiment by Neumann. Detailed calculations are presented on the role of free oxygen in axial transport and burnout of the lamps.

Paper 7

HEAT AND MASS TRANSPORT IN GAS-FILLED INCANDESCENT LAMPS

J Fitzgerald and E Fischer (Philips, Aachen, Germany)

The problem of determining heat and material transport from incandescent filaments in gas-filled lamps is treated theroretically and the results experimentally verified. Unlike the semi-empirical approach of Langmuir [1] and Elenbaas [2], [3] the model [4] utilizes the complete set of equations of fluid motion and enables the total transport and associated fields, i.e., the temperature and flow-field of the inert gas and the tungsten flow pattern to be calculated. For the sake of both experimental and analytical simplicity attention has been confined to a horizontal filament with a coaxial cylindrical envelope which can be treated as a two dimensial problem.

The analysis proceeds from the equations of fluid dynamics, i.e. mass-balance, energy-balance and momentum-balance (Navier-Stokes equations), and the diffusion-equation. These are tailored to suit conditions prevalent in gas-filled lamps, ie., relatively small inertial forces, moderate gas-flow velocities and extremely dilute conditions for the binary gas mixture of tungsten and inert gas. These simplifications enable the Navier-Stokes equation to be expressed as an interaction between two forces, the buoyant and viscous forces. A suitable stream function is introduced to satisfy the mass-balance: the remaining problem is with the energy and momentum-equations which are coupled by the temperature and velocity fields.

These are solved numerically by iteration; the first approximation for the temperature field being obtained from the convection free energy balance. The spatial distributions temperature and velocity obtained by this procedure are used to determine the tungsten and heat flow fields and hence the total losses.

Experimental verification is obtained by comparison with temperature and gas flow fields measured by holographic interferometry [5], [6] and a particle tracer technique, respectively. The effect of pressure variation is used to demonstrate the good agreement between theory and experiment. It is found that the envelope diameter strongly influences the transport and an optimum diameter exists at which both the heat and mass exhibit a minimum. Further experimental studies show that the wire position also was a strong influence on the temperature and flow fields.

[1] Langmuir, I., Phys. Rev., 401 (1912)

[2] Elenbaas, W., Physica 4, 761 (1937)

[3] Elenbaas, W., Physica 6, 380 (1939)

[4] Fischer, E., Fitzgerald, J., J. Appl.Phys. 45, 2895 (1974)

[5] Fitzgerald, J., Horster, H., Philips Tech.Rev., 32, 206 (1971)

[6] Phillips, N. J. Coaton, J. R., Lighting Research and Technology 5, 156 (1973)

Paper 8

PROJECTION HOLOGRAPHIC INTERFEROMETRY AS ATECHNIQUE FOR THE DIAGNOSTICS OF GAS-FILLED LAMPS

H J Phillips (University of Technology, Loughborough, Leics)

The technique of producing double-exposure holograms as a laboratory diagnostic method in the examination of lamps, or other objects which are of a partially transparent nature, is now well established. Workers in this field will have been impressed by the power of the method in the production of clear impressions of interference patterns as a result of redistribution of the internal gas when the lamp is heated by its own filament.

The most severe mathematical complexity is associated with the interpretation of the observed interference patterns and a direct method of analysis would seem to be highly desirable.

In this paper a discussion is given of the method of projection holography whereby a projected image of the lamp is produced holographically in either single or double-exposure modes.

Some fascinating possibilities can be demonstrated namely the diagnostics of the interiors of visibly inaccessible objects and the direct readout of interference patterns into a computational system.

Paper 9

GAS-PHASE SEPARATION MECHANISMS IN LINEAR TUNGSTEN-HALOGEN LAMPS

J R Coaton (Thorn Lighting Ltd, Melton Road, Leicester)

It is well known that the burning position of long, linear tungsten-iodine lamps is restricted to a few degrees from the horizontal, because of halogen separation within the gasfilling. This causes the tungsten transport cycle to breakdown at the upper end of the lamp, leading to local bulb blackening. Experimental and theorectical work has shown that the separation mechanism can be adequately described by the Furry, Jones and Onsager model for isotope separation in a Clusius and Dickel column, modified to take account of angle burning operation. The results reveal that there exists a critical burning angle for every gas-filling content and pressure where the degree of halogen separation reaches a maximum, and this behaviour may be explained by proposing the existance of a convection cell within the lamp, which is retained at all burning angles. This has enabled a critereon to be established for the design of genuine universal burning linear tungsten-halogen lamps, capable of operating at any angle, without bulb blackening and/or erosion of the solid tungsten in any part of the lamp. It also clearly shows that whilst high pressure gasfilling may avoid halogen separation in vertical burning lamps, it does not necessarily produce universal burning characteristics. In fact, the criterion for universal burning is difficult to satisfy, but successful experimental lamps have been produced, gasfilled with xenon and containing bromophosphonitrile.

Further support for the proposed searation has been secured by dynamically sampling the gaseous contents of linear lamps running at various angles. Here, the results show that nitrogen does separate in argon and krypton gasfillings, and separation almost certainly occurs where any gaseous species differs in mass and bond energy from the major gasfilling component.

The possibility of halogen separation caused by operating linear lamps from a direct current sup.ly has been briefly considered - and dismissed.

Paper 12

TEMPER..TURE AND VELOCITY FIELD BETWEEN COAXIAL HORIZONTAL CYLINDERS IN FREE CONVECTION

A Szokefalvi-Nagy, G Tichy, I Gaal (Research Institute for Technical Physics of the Hungarian Acedemy of Sciences, Budapest, Hungary)

The hydrodinamical equations for free compressibile gas convection will be solved by different methods between coaxial horizontal cylinders. The temperature of the inner cylinder /modeling the incandescent filament/ is homogeneous and the radial temperature distribution on the outer cylinder /bulb/ in prescribed. It turns out that the amount of the net convective heat transfer is governed by the asymmetry of the velocity field with respect to the horizontal geometrical symmetry plane of the arrangement. Therefore the net convective heat transfer can vanish/i.e. the Langmuir layer can reach the bulb/ also in cases in which the symmetrical part of the velocity field /giving rise to symmetrical kidney shaped flow lines/ is appreciable and the convective mass transfer from the filament is important and asymmetrical. By means of some diagrams an impression will be given on the velocity and temperature distribution in the tube at different parameters characteristic for thin tubulare tungsten halogen lamps. Especially the role of the temperature dependence of the vyscosity and of the inhomogeneous wall temperature will be discussed. The heat and mass transfer will be described also in terms of the similarity parameters of the hydrodinamical equations.

Paper 13

SOME PROBLEMS OF FORMING THE IMAGE OF GASEOUS (PHASE) OBJECTS

J H Toyer, N J Phillips (Dept of Physics, University of Technology, Loughborough)

It is now well known that considerable information about the state of the interior of a gas-filled lamp can be obtained by interferometric methods of which the double-exposure holographic technique would seem to be the most important.

The impressive difficulty of interpretation of the fring patterns observed in this or other methods (e.g. Mach-Zender interferometry) may perhaps be alleviated by using an optical imageing technique so as to exploit depth of field methods.

For example, if aview is obtained of a cylinder heated by an axial filament using an imageing lens system, it would be pertinent to enquire whether such a phase medium is subject to the same rules as ordinary objects viewed in incoherent light. If so, looking at a small region of it bounded by planes of the limits of "depth of field" should be possible.

This problem has not been explicitly solved before and in this paper, we look at some ideas which at least help us to understand the difficulties of the method.

It is indicated that a slice of phase medium can be represented by a diffraction grating of a general type whose spatial frequency spectrum is related to the physical properties of the medium involved. Once this step is taken, it is a simple progression to discuss the production of an image of such a generalised grating and its possible "depth of field" properties.

It will be indicated that extension of the ideas to work in Mach-Zender interferometry or Schlieren studies are potentially useful.

Paper 14

SPECTROSCOPIC TECHNIQUES FOR DIAGNOSTICS OF METAL HALIDE ARC LAMPS

J F Waymouth, W K Keeffe, W I Bamberg, P J Gardner, W Walter, D Work
(GTE Sylvania Incorporated, 100 Endicott Street, Danvers, Mass, U.S.A)

Progress in understanding the complex chemical and physical processes which occur in metal halide lamps has been hampered in the past by lack of the ability to measure concentrations of the important species in the arc column. We have developed spectroscopic techniques for the measurement of arc temperature and the concentrations of important atomic and ionic species in the arc, which are yielding important information for our lamp development and improvement program.

These are based on measurements of two types: Measurements of spectral radiance of selected optically-thick lines, for determination of axis temperature of the arc by means of two-line Bartel's method; and measurements of spatially-integrated flux emitted from a short segment of the arc column in selected optically-thin lines for determination of axis temperatures and concentrations of the several species. All measurements are phase-resolved, customarily at the phase angle of maximum temperature of ac-operated arcs.

Reduction of spectral flux and radiance measurements to the desired temperatures and concentrations is carried out assuming parabolic-temperature-profile or two-temperature effective-temperature, effective-radius models. Agreement between Bartels-method (independent of temperature profile) and optically-thin spectral-flux measurements of arc axis temperature (assuming parabolic temperature profile) is excellent.

Axis temperatures of metal halide lamps containing mercury, scandium iodide, and sodium iodide are found to be typically 5500°K at phase angle of maximum temperature. The total concentrations of scandium and sodium atomic plus ionic species are in reasonable agreement with the vapor pressures of scandium iodide and sodium iodide at the wall temperatures, and disagree markedly with the prediction from Knudsen-cell measurements that the predominat vaporizing species is $NaScI_4$.

Paper 15

EMISSION LINES IN THE SPECTRA OF HALOGEN INCANDESCENT LAMPS

W J van den Hoek and E G Berns (Philips, Eindhoven, Netherlands)

The spectral emission of a number of halogen incandescent lamps consists of a line-emission spectrum superimposed on the continuous thermal emission from the tungsten coil. These emission peaks come from atoms in the gas phase, which were originally present in the coil. Spectra from W, Mo, Fe, Al, Ti, V and Cr are observed. The composition of the line spectrum depends on the type of filling gas and on the origin of the tungsten filament. The mechanism and the consequences of the observed phenomena are discussed.

Paper 16

SYNCHROTON RADIATION SOURCES

I H Munro (Daresbury Laboratory, UKAEA, Warrington, Lancs)

At Daresbury, the electron synchrotron "NINA" is utilised as a source of synchrotron radiation, covering the visible, ultra-violet, and x-ray spectral regions. Several university-originated research programmes are in progress at the synchrotron radiation facility here, encompassing such fields as atomic and solid state physics, metallurgy and biology. Additionally, the National Physical Laboratory is using synchrotron radiation to calibrate transfer standard sources in the ultra-violet.

Paper 17

AUGER ELECTRON SPECTROSCOPY AN ELECTRON SPECTROSCOPY FOR CHEMICAL ANALYSIS

M J B Thomas (GTE Sylvania Incorporated, 100 Endicott Street, Danvers, Massachusetts, U.S.A.)

Rapid development in availability of instrumentation has facilitated the inclusion of Auger Electron Spectroscopy and Electron Spectroscopy for Chemical Analysis (ESCA) as members of the array of surface analysis techniques available for diagnostics. The ability to look to a depth of no more than 10 to 25 angstroms and on a fairly routine type analysis has made these approaches a requirement not only for the acedemia, for the R and D laboratory, but for production-type problems. When applied to various incoherent lamp sources, problem sources, or reactivities have been revealed which heretofore would have been incapable of analysis.

The paper will be roughly divided into three sections: I-Use of AES and ESCA for the determination and/or source of various visible phenomena observed in the production or during the operation of such of tungsten-halogen or quartz arc tube lamp types. One of the specific examples in this category is correlation of discoloration produced on arc tubes in lamp processing with the amount of reducing atmosphere present and the valence state of silicon in the quartz envelope.

II-Correlation of surface effects detected with variable performance of raw materials used for lamp production. As for example, presence of residue surface material from a raw material process wash caused by the inability of the material to be soldered.

III-Identification of valence variations within materials used in incoherent lamp sources and correlation with actual performance. Examples in this category will be related to production of surface layers during operation of quartz iodine lamps, which lead to distinct discolorations related only to valence change.

Paper 18

AN APPROXIMATE IMAGE EQUATION FOR A NEW METHOD OF THE OPTICAL DIAGNOSTICS OF A GLOW DISCHARGE POSITIVE COLUMN

A Sprachta (Tesla VUST, Prague, Czechoslovakia)

A theoretical basis of a new way of measuring the atomic concentration in different levels in a glow discharge positive column which may be considered a non-coherent three-dimensional radiator has been described in [1] and with more details in [2].

The image equation for the experiment description is very extensive and difficult to analyse. This was a reason why an approximate image equation was made to give much better summary of the experiment arrangement possibilities and the method precision.

The image equation has been newly devised because of the fact that in the image plane neighborourhood with the points corresponding to the lens equation the optical system impulse response is equal to the Dirac δ - function.

A real experiemnt arrangement has been included in my work. As a base of the experimental arrangement optimalization the contractive mapping condition for an approximate equation is used.

The simplified conditions might be used to measure the relative atomic concentration.

The results of relative atom concentration measurements in Argon and Neon may be presented.

[1] Sprachta A.: Proc.XI-th ICPIG, Prague 1973
[2] Sprachta A.: Czech.J.Phys.B 24 (1974), 366

Paper 19

PHOTONPRODUCTION AND INTENSITY CORRELATIONS

A A Walma (Dept of Electronic and Electrical Engineering, University Sheffield, Sheffield)

The photonproduction within a light source is treated as a Markov-chain. Denoting the state of the process (which is the number of photons at a certain time) by the random variable x(t), the characteristic function $C_t(\phi)$ can be shown to satisfy [1]:

$$\frac{\partial C_t(\phi)}{\partial t} = \psi\left(i\phi, t, \frac{\partial}{\partial i\phi}\right)C_t(\phi) \qquad \ldots(1)$$

where the operator $\partial/\partial i\phi$ acts only on $C_t(\phi)$ and is ψ determined by the processes involved. With a suitable formulation of the problem, the appropriate partial differential equation (1) can be written down more or less at first sight. In applying (1), the derivation of the master equation is circumvented. It should be noted that the solution of the master equation in terms of its characteristic function, results in a formula of the kind of (1).

Euation (1) will be applied to a number of light-sources. A single-photon Townsend discharge can be shown to possess a 'bunching' output. An explicit geometric distribution will be derived for a general class of lasers. For some lasers (e.g. Ne-He), discharge sources (e.g. low pressure Hg) and electroluminescence devices (e.g. GaAs laser) Poisson output statistics will be derived, with a mean depending on the underlying processes. In the cases where equation (1) is intractable, the second moment of the number of photons can still be obtained for a characterisation of its fluctuation properties.

[1] M S Bartlett, An Introduction to Stochastic Processes, Cambridge University Press, 1955

Paper 20

A CONSISTENT ANALYSIS OF INCOHERENT RADIATION TRANSFER PROCESSES IN DISCHARGE LIGHT SOURCES

M A Cayless (Thorn Lighting Limited, Leicester)

Recent analyses of discharge light source plasmas in non-LTE or near LTE conditions have made use of the Biberman-Holstein treatment of resonance radiation transfer. Attempts to use a similar method for high intensity discharges containing several reacting elements or molecules reveal a number of inadequacies: it is not suitable for non-resonance line, band or continuum radiation; it is only valid for a plasma with spacially constant absorption coefficients. Attempts to modify the treatment to fit these conditions produce results which are inconsistant with general radiation theory, needing special analytical techniques for their evaluation.

On the other hand, most published analyses of the plasmas of high pressure discharges in mixed reacting vapours do not incorporate a satisfactory treatment of the effects of radiation production and transfer on the plasma properties. These can have a marked effect on such features as temperature profiles, spacial variations in spectral composition and component separation, all of which are of considerable interest in the development of satisfactory light sources.

In this paper the general theory of radiation transfer is applied directly to the analysis of these discharges in conditions in which LTE may not be fully established. The key feature is the evaluation of the spectral photon concentration as a function of position. The result is a rather modest generalisation of the Biberman-Holstein treatment which has two great advantages: it is not an extension of a more specialised treatment into situations for which it was not derived; and it is fully consistent with general radiation transfer theory, so that the numerous computational techniques which have been developed for a wide variety of applications are available to help find satisfactory solutions. The relations reduce to the Biberman-Holstein form in the special case of line radiation in a uniformly absorbing plasma.

Paper 21

HALIDE COMPLEXES IN HIGH-PRESSURE DISCHARGE LAMPS

A G Chalmers, D O Wharmby, F L Whittaker (Thorn Lighting Ltd, Leicester)

The range of elements excited in high pressure discharge lamps was greatly extended from the traditional mercury, sodium and the permanent gases by the advent of the metal halide lamp. Although the vapour pressures of simple halides are sufficient for a great number of elements to enter the arc under suitable conditions, there remain some elements whose halides are relatively involatile. The halides of sodium are less volatile than the elementary metal and sodium chloride, for example, is not an effective means of introducing sodium into the arc of conventional metal halide lamps, where the cool-spot temperature is typically 500-800°C.

This paper describes how it has been found that when sodium chloride is added to an arc tube along with other halides, much more sodium radiation is emitted than would be expected from the sodium chloride alone, or any sodium impurities. Subsequent experiments and calculations show that this result can be attributed to the formation of halide complexes in the dose condensate. For example, sodium chloride combines with aluminium chloride to form $NaAlCl_4$ and with stannous chloride to form $NaSnCl_3$. These compounds vaporize as $NaAlCl_4$ and $NaSnCl_3$ with appreciable partial pressures. Such complexes may provide a means of exciting a yet wider range of involatile atoms in halide arcs. The formation of complexes of definite composition may also promote a more constant spectrum and colour throughout the life of the lamp, by ensuring that the concentrations of the emitting species are in a constant ratio.

Paper 22

ENHANCEMENT OF THE PARTIAL PRESSURE OF LOW VOLATILITITY HALIDES IN HIGH PRESSURE DISCHARGE LAMPS

R Lorentz (Philips, Eindhoven, Netherlands)

Metal halides used in high pressure gas discharge lamps can be divided into two groups with respect to thir volatility.
One group consists of covalent salts with sufficient high volatility so that the optimum density of the light emitting species can be reached by complete vaporization of the compound.
In the other group one will find the ionic salts with low volatility where the vapour pressure and thus the density is determined by the wall temperature, for example the halides of the alkali metals, the aldali earth and the rare earth group. In this case enhancement of the wall temperature will lead to a higher partial pressure and to a higher luminous efficacy. This possibility is limited by the termal and chemical properties of the wall materials.

Other parameters, which can influence the vapour pressure, such as the inert gas pressure and the particle size of the condensed phase, result in only a small increase in vapour pressure.
From recent literature it is known that the effective partial pressure of low volatile halides can be increased due to the formation of gaseous complexes. It has been found that this effect can be successfully applied in high pressure gas discharge lamps.
The bound structure and the thermodynamic properties of these complexes (or heterodimers) are comparable to the properties of the better known pure dimers (or homoeodimers).
For the application under high pressure gas discharge conditions, some selection rules can be derived in order to combine suitable metal halides which lead to complexes stable enough at higher wall temperatures.
On one hand it has been found that the complexes are most stable when the combining compounds also form stable dimers; on the other hand one should choose a halide with a high volatility as the complex forming partner.
Lamp experiments are reported for lamps with sodium halides and the halides of the alkali earth group, where the best results could be obtained with aluminium halides as the complex forming partner.
The effective partial pressure of the low volatile halide could be enhanced to such an extent that the luminous efficacy increased up to 50% , compared to the values obtained without complex formation.
The use of these stable gaseous complexes gives a new parameter for the development of metal halide lamps, offering the possibility of altering the partial pressure of the light emitting species to a considerable extent independently of the wall temperature.

Paper 23

EFFECTS OF MASS TRANSPORT ON ENERGY BALANCE IN METAL HALIDE ARCS

G J Kazek (General Electric Co., Cleveland, U.S.A)

Wall temperature distributions have been measured for vertically operated arc tubes which contain rare earth halides, sodium iodide and mercury. The arc tubes were constructed such that the halide condensate was forced onto the inner cylindrical surface of the tubes rather than being permitted to remain in the end chambers. Results indicate that geometry, input wattage, and mercury mass are the most important parameters affecting the average arc tube wall temperatures. On the other hand, the dominant factors controlling the shape of the arc temperature distribution are the total mass of halide and the halide mole fractions. Observations of the sodium distribution in the arc and arc spectra obtained at various axial positions reveal that the energy balance within the arc varies considerably (in the axial direction) as the halide parameters are altered. The roles of convection and mass transport processes in arc energy balance will be briefly examined in order to give some insight into the consequences of the foregoing data on theoretical modelling.

Paper 24

AN ELEMENTARY ARC MODEL OF THE HIGH PRESSURE SODIUM LAMP

J F Waymouth (GTE Sylvania Incorporated, Danvers, Massachusetts, USA)

An elementary arc model of a high pressure sodium lamp based on a two-temperature effective-radius concept has been developed. Calculations of the radiant power and luminous output emitted in resonance-broadened sodium-D lines, the radiant power emitted in infrared and visible lines, and heat conduction loss as a function of sodium vapor pressure are in good agreement with measured values assuming measured values of axis temperatures, for reasonable values of effective radius.

Complete accounting for all of the observed radiation power, and satisfaction of the energy balance, require the assumption of a continuous spectrum emission proportional to the square of the sodium vapor pressure. Possible sources of such radiation are emission from Na_2 molecules or NaHg excimers. Several formalisms for independently defining the effective radius as part of the calculations are presently under consideration.

The semi-empirical two-temperature model offers a distinct advantage over first-principles calculations for engineering design of high pressure sodium lamps, provided it can be refined to sufficient accuracy. The complete calculation of all observable quantities for given lamp geometry, sodium and mercury vapor pressures, and assumed arc temperature requires only two seconds on a computer of modest capability. First-principles calculations involving solutions of the Elenbass-Heller equation including optically-thick and optically-thin radiation as well as heat-conduction loss require orders of magnitude more computational power.

Paper 25

MEASUREMENTS OF ELECTRODE POWER DISSIPATION IN HIGH PRESSURE SODIUM LAMPS

P L Denbigh, D O Wharmby (Thorn Lighting Ltd, Leicester)

In low-power high-pressure sodium lamps the power dissipated in the electrode regions forms a substantial fraction of the input power. Since comparatively little light is produced in the electrode regions, it is important to establish methods for comparing the merits of various electrode designs with a view to making improvements.

We have used short synthetic sapphire arc tubes 7.6 mm bore dosed with sodium amalgam and xenon. Electrode tip separations varying from 0.6mm upwards have been used. The arc tubes were operated on a 50Hz supply at constant currents between 1A and 5A with one end submerged in liquid indium to control the amalgam temperature.

The arc tube voltage was measured as a function of amalgam temperature and in every case we observed a voltage minimum between amalgam temperatures of 300 and 400°C. The voltage rise at higher amalgam temperatures is consistent with the known values of the electric field in the high-pressure sodium arc. Comparison of the minimum voltages with electrode voltages estimated from experiments with lamps of different length indicates that the minimum is related to the electrode voltage drop from which the electrode power dissipation is estimated. The voltage minimum varies with electrode separation indicating that the discharge has a fully developed positive column for separations exceeding 15mm. The voltage minimum observed for pure tungsten electrodes was usually twice that found when the electrode was coated with emitter. At a current of 1A the voltage minimum was typically 30% higher than at 4.5A.

Paper 26

THE TIN HALIDE ARC

P C Drop, J J de Groot and A G Jack (Philips' Gloeilampenfabrieken, Light Division, Eindhoven, Netherlands)

When exploiting radiation generated by molecules, light sources with an excellent colour rendition can be made. A good example of this is the mercury - tin iodide - tin chloride lamp. The major part of the emitted continuum radiation is due to tin monohalide molecules. Adjacent to the wall a sheath of tin diiodide molecules gives rise to a strong absorption of radiation, especially at shorter wavelengths. This has a strong influence on the spectrum and can be used to change the colour temperature within a wide range. It is found that the amount of chloride has no influence on the colour temperature because the absorption by tin dichloride is negligible. In a properly designed lamp all the constituents are vapourized, i.e. the spectrum is independent of input power.

Some of the problems of this kind of lamp will be discussed. First of all, in a vertically burning, wall stabilized lamp, instabilities occur at high tin halide pressures. Secondly, when burnt horizontally the arc bows upwards, causing a local overheating of the quartz tube wall. This problem can be solved by applying a relatively small magnetic field to a horizontally burning lamp, making use of a special construction of the inlead wires. The temperatureof the upper section of the discharge tube now becomes rather insensitive to changes in the input power. Also there is now no upper limit to the tin halide pressure.

One other problem is the fact that, although the colour rendition is very good, the colour appearance is greenish. This can be overcome bt the addition of Li or In or both. Although unlike the other constitituents the Li is not completely vaporized, the spectrum remains insensitive to changes in the input power.

Results obtained with experimental lamps will be presented.

CW AND PULSED CADMIUM AND ZINC METAL VAPOR ARC LAMPS AS EFFICIENT LIGHT SOURCES FOR DEEP OCEAN WATERS

M Geller, D E Altman and G J Barstow
Naval Electronics Laboratory Center, San Diego, California 92152

The optical transmission characteristics of deep ocean waters show a broad maximum between 4600 Å and 5000 Å.[1] Optical sources for underwater applications should emit in this ocean window, or close to it, to minimize the required optical power.

There are few reasonably efficient high power, CW optical sources available in this range.[2] The argon ion laser, emitting 4-5 watts at 4880 Å and 5150 Å is the best source currently available for underwater use. This laser is limited to an efficiency of less than 0.05% Furthermore, for underwater systems, the coherence properties are of little value because of the large amount of scattering.[3] Among incoherent sources, thallium iodide long arc lamps radiating tens of watts in the broadened 6^2P-7^2S transition (5350 Å) of thallium have allso been used. However, 5350 Å is not at the optimum wavelength for deep ocean waters. The only incoherent source emitting appreciable power in this spectral region is the high pressure zenon short arc lamp.[4] As its output in the visible issimilar to that of a black body, the efficiency of radiation into the 4600 - 5000 Å band is low.

A pair of new metal vapor short arc lamps to fulfil this need for an efficient CW light source for deep ocean waters has been developed.[5] The dominant emission from cadmium is in the 5^3P-6^3S transition at 4678, 4800 and 5085 Å, and from zinc in the 4^3P-5^3S transition at 4680, 4722 and 4811 Å. Efficiencies of conversion from electrical to radiant energy into these triplets was as high as 5% with usable powers of tens of watts. The techniques of lamp processing for operating lifetimes of hundreds of hours were developed.

The most commonly used pulsed source for underwater systems is the frequency doubled Nd^{3+} laser radiation at 5300 Å. However, the attenuation coefficient at this wavelength is about twice as large as the minimum value at 4800 Å. There is a good potential of developing a convenient, inexpensive, pulsed source for underwater systems as kilowatts of peak power were obtained from a pulsed zinc short arc lamp. Time resolved spectra from electrically pulsed zinc vapor discharge show a strong peak power emission in the blue-green: the triplet $^3P-^3S$ that was dominant in the CW spectrum and additionally a $4f^2F-4d^2D$ doublet of the ionized spectrum at 4912 and 4924 Å. The time histories of the ground states of the triplet levels of ZnI and the doublet levels of the ZnII were measured during the electrical discharge. Unexpected high population densities of the excited states were obtained.

References:

1. D E Matlock, "The Deep Ocean Optical Measurement (DOOM) Program", N.O.L. Report TR-70-165, Whiteoak, Md.

(further references overleaf)

2. R J Pressley, Ed, Handbook of Lasers with Selected Data on Optical Technology, (Chemical Rubber Co, Cleveland Ohio, 1971). In Sec. 2, there is a comprehensive review of Incoherent Optical Sources by I Leiberman.

3. C F Gallo, Appl. Opt. 6, 1563 (1967); 9, 2711 (1970).

4. V H Schirmer, Z Angew. Phys. 11, 357 (1959).

5. M Geller, D E Altman, G J Barstow, Appl. Optics 11, 1439 (1972).

6. C H Corliss and W R Bozman, "Experimental Trasition Probabilities for Spectral Lines of Seventy Elements", Nat. Bur. Stand. Monograph 53 (Government Printing Office, Washington D.C., 1962).

Paper 28

A NEW, COMPACT FLUORESCENT TUBE

C E Williams (GEC, Hirst Research Centre, Wembley)

The linear form of fluorescent tubes in the standard range imposes aesthetic restrictions on architects' designs for building interiors. As a result of this a demand has arisen for a light source suitable for use in a square modular fitting of approximately 0.6 metres side.

The achievement of the requisite light output by the use of a multiplicity of small standard fluorescent tubes in each fitting is unattractive for a number of reasons, including the basic one of unfavourable distribution of energy between electrodes and positive column. The ideal arrangements in this respect is a single discharge tube with a long positive column and only a single pair of electrodes.

Such a lamp, if of sufficiently small diameter, may be bent into a grid form to enable it to be accommodated in the luminaire. The specialized manufacturing processes thus required entail a substantial cost which can only be justified if an exceptionally long lamp life is achieved. This is made possible by the use of electrodes of robust construction, together with an inert-gas lamp-filling at a pressure considerably in excess of that customarily used in orthodox mains voltage fluorescent tubes, where it would be unacceptable because of the low efficiency resulting from increased electronic energy losses by elastic collisions. It is tolerable in the new design of lamp because these losses are off-set by the reduction in the proportion of input energy dissipated at the electrodes. Additional advantages of the high pressure arise from improved total lumen output and lumen maintenance.

The new lamp has an operating voltage of 415 volts, and is conveniently supplied from a leakage reactance Transformer having sufficient open circuit voltage to ignite the discharge without cathode preheating. The tube wattage is 153, the luminous flux 7860 lumens and mean life 20,000 hours.

Paper 29

MULTICHIP SEMICONDUCTOR LAMPS

R Hall, J R Fitzpatrick, B F Jones (Thorn Lighting Ltd, Leicester)

Further development of the present range of semiconductor lamps is envisaged by the use of two or more electroluminescent chips per package. The type of lamp and its application will depend upon whether the chips are connected (a) in series or (b) in parallel. The use of series connected chips implies that the chips are excited simultaneously producing a high forward voltage lamp, whereas for a parallel connection normally only one chip would be energized at one time.

Using both red and green emitting gallium phosphide, prototype lamps have been produced in which up to four chips have been stacked in series. Measurement of the electrical characteristics and the total capacity have confirmed that the series resistance is small. The increase in light output obtained by using 2 and 3 chips as compared to the single chip lamp has been found to be a factor of 1.7 and 2.4 times approximately for the geometry used.

It is expected, that for the 'high-voltage' lamp, the chips will normally be chosen to emit the same colour, whereas the most probable development for the parallel connected chips will be the dual or mult-colour lamp. The possibility of fabricating electroluminescent chips in the p/n or n/p mode increases the number of configurations which are possible.

Dual colour lamps have been fabricated using 2 pin and 3 pin bases. The compatibility of the electrical configuration for different circuits is emphasized. In the fabrication of the dual colour lamps the chips are normally placed as close to the optical axis as possible. The effect on the viewing angle of this small displacement is dicussed.

Paper 30

DEEP LEVEL LIMITED EFFICIENCIES IN GALLIUM PHOSPHIDE LED'S

B Hamilton, T J Hayes, A R Peaker and B L Smith (Ferranti Ltd, Chadderton, Oldham, Lancs)

Although very large improvements in the performance of light emitting diodes have been achieved in the last five years the overall energy conversion is still very low in absolute terms particularly for green emitting devices (~0.1%). This paper traces the reasons for this low effeciency and shows that the principle energy loss mechanism in green emitting gallium phosphide devices is a non-radiative recombination process basic to the bulk material. Other potentially limiting factors such as carrier injection efficiency and light extraction are important but can in the main be controlled by existing technology.

Numerous potential non-radiative recombination mechanisms exist, Auger band-band processes may set an effective limit to the lifetime and radiative efficiency at high doping and injection levels. At moderate doping and injection levels normally met with in indirect band gap LEDs, the poor radiative efficiency is more likely to result from non-radiative recombination at states deep in the forbidden energy gap. The influence of such states on material parameters has been difficult to assess, because the states themselves are virtually impossible to detect in bulk material when they are present in the relatively low concentrations which limit the minority carrier lifetime. Recently, however, powerful techniques have been developed which enable the detection of minute concentrations of deep states provided they are located in the depletion region of a p-n junction or Schottky barrier. Using these experimental methods it has been possible for the first time to make a definitive study of the effect of traps on luminescence in the III-V compounds.

For the case of green emitting gallium phosphide it is shown in the paper that the minority carrier lifetime and the luminescent efficiency are limited by the presence of deep states and the significance of this in relation to future improvements in the efficiency of green light emitting diodes is discussed.

Paper 31

TRANSIENT BEHAVIOUR OF LIGHT EMITTING DIODES

A Descombes (Federal Institute of Technology, Zurich)

In this paper, the results of experimental investigations of the transient behaviour of light emitting diodes are presented. The influence of diode current density, bias voltage and temperatue on the characteristic response time are discussed. The devices investigated were Si-doped GaAlAs infrared emitting diodes. The results are compared with visible emitting GaP diodes. In a brief summary, the problems arising with the measurements of fast light pulses are discussed. A tentative theoretical explanation of the results is presented.

One of the most promising applications of GaAs infrared emitting diodes is optical communication for short distances, either directly through the air or through optical fibres. CW operation at room temperature is possible with good reliability, such as that of commonly used transistors. Visible emitting GaP or GaAsP diodes are mainly used as display devices.

In both cases, the knowledge of the transient behaviour is very important. In the case of optical communication, light rise and fall times as well as delay times between current and light pulses are factors limiting transmission capacity. In the case of display devices, light decay times or say fall times can be used to determine quantum efficiencies. Calculations of minority carrier lifetimes from eithertransient measurements or quantum efficiency measurements often led to different results. That means more work has to be done in order to find a more exact relation between transient parameters and external quantum efficiency, current density, device temperature and geometrical configuration of the devices. That's the main aim of this work. A new method of analyzing transient curve traces from a sampling scope is used which allows the determination of the characteristic switching times with an accuracy greater than that of other methods, for instance the photographic one.

In the first section I shall show briefly some typical structures of light emitting diodes and some few typical features. In the second section I shall describe the experimental set up used for measuring transient parameters[1] and
some problems arising with the measurement of fast light pulses. 2
In a third section I shall describe and discuss the results of the measurements. In the last section I shall give tentative theoretical interpretations of the measurement results and some conclusions drawn from them.

Note 1 - Choice of the right detector
Checking response time of the detector
Nanosecond pulse problems
Evaluation software

Note 2 - switching times vs diode current
switching times vs temperature
influence of Dc bias voltage
comparisons of GaAs and GaP diodes

Paper 32

FACTORS AFFECTING EFFICIENCY OF POWDER ZnS DCEL DISPLAYS

P H J Beatty (GEC Hirst Research Centre, Wembley)

Variables affecting power efficiency in powder ZnS:Mn, Cu, Cl DCEL displays include particle size, copper in the lattice and surface of particles, method of delineating the transparent positive electrode, encapsulation atmosphere and electrical operating conditions.

Small average particle size is less efficient although brighter. This is possibly the result of higher currents obtained with smaller particles leading to a greater thermal danage.

With 0. 1 %wt. Cu incorporated in the ZnS lattice during firing of the phosphor, a brighter panel can be obtained, although the current is not altered significantly. At low current values, where other factors are not involved, such as Cu_xS precipitation, efficiency can be up to twice as great as for Cu free material. Since the Cu doped material tends to contain more hexagonal ZnS, it is possible that the improved luminance is the result of a different local environment of Mn^{2+} luminescent centres within the lattice.

When voltage is first applied to a panel an electrical forming occurs in which diffusion of copper ions may take place to produce a barrier for injection of carriers. This gives a more resistant layer of particles immediately adjacent to the positive transparent electrode (SnO_2-coated glass). It is necessary to further delinate this electrode with a thin layer of insulator to prevent leakage current via low resistance unformed phosphor next to the area required to be lit. If this procedure is not used, the unformed areas may eventually form and light gradually spread out a few millimetres, or the efficiency is gradually reduced still further since the current remains high whilst the luminance falls to a somewhat lower value after the initial forming and running-in period.

It is particularly important to exclude water vapour from the encapsulation atmosphere since it can decrease efficiency by raising current more than luminance. This is thought to occur by a charge transfer adsorption process on the surface of ZnS. Likewise, a similar mechanism explains why O_2 can lower current more than luminance and increase efficiency. Panels operated on short pulses, in an inert atmosphere or vacuum, show steady luminance and current at low efficiency, but in oxygen there is improved efficiency.

For a panel which has been operated at 100 V DC, both luminance and efficiency increase up to 100 V DC. Sometimes, efficiency reaches a maximum at the highest voltages, limited possibly since power dissipation thermally destroys the light emitting part of the phosphor whilst still permitting current to flow.

Under DC pulse operation, efficiency rises during life of the device as current falls more than luminance. Efficiency can be higher than with continuous DC, especially for short pulses in an operating atomosphere containing O_2.

Thus, there is a choice of preparative techniques for phosphor and panel preparations and a variety of operating conditions depending on requirements for high luminance or efficiency with very high efficiency possible at a somewhat lower brightness.

Paper 33

THE MICROSTRUCTURAL AND MECHANICAL BEHAVIOUR OF TUNGSTEN WIRES AT ELEVATED TEMPERATURES

J H Day (Thorn Lighting Ltd, Melton Road, Leicester LE4 7PD)

The microstructural features that develop during the high temperature ageing of tungsten wires used in the manufacture of lamp filaments have been examined by optical and electron microscopy. The grades of tungsten investigated include pure tungsten, tungsten doped with 0.6 %thoria and tungsten doped with aluminium, potassium and silicon (AKS). In the doped tungsten the dispersoids are responsible for the elongated grain structures formed during recrystallisation through a preferential grain boundary pinning action. Grain boundary migrations are inhibited by oxide particles in the thoriated tungsten and by linear arrays of voids in the AKS tungsten. The high strength of the dopant-containing tungsten can be largely attributed to the elongated grain structuresbut it is not clearly understood what contributions, if any, the dispersoids themselves make to the strength at elevated temperatures. In the present work an attempt has been made to obtain a better understanding of the factors that control the resistance to deformation through an investigation of the high temperature creep properties of the different grades of tungsten.

Constant-load creep tests have been performed under vacuum conditions at temperatures between 1800 and 2500°C using specimens having a helical coil geometry. It was found that the stress dependence of the creep rate varied with the test temperature and stress and that the activation energy for creep varied between 135 and 180 k cal per mol. The stress dependence of the pure tungsten varied between 3 and 5 with somewhat higher values for the thoriated tungsten. For the AKS tungsten the values of the stress exponent increased with temperature and were found to vary between 12 and 38. Such high values of the exponent are characteristic of some dispersoid strengthened systems and it is therefore suggested that the finely dispersed voids in AKS tungsten not only control the recrystallised grain morphology but contribute to high tempeature strength by impeding the movement of dislocations.

Paper 34

CARBON AS AN ALTERNATIVE FILAMENT MATERIAL

W J van den Hoek (Philips, Eindhoven, Netherlands)

A discussion will be given on the properties of graphitic carbons which have relevance to the application of carbon in incandescent lamps. These properties include: graphitization, mass transport and chemical transport.

Paper 35

THE EFFECT OF POROSITY AND GRAIN STRUCUTRE ON THE LIFETIME OF INCANDESCENT LAMP FILAMENTS

O Horacsek, K Horacsek, L Bartha (Research Institute for Technical Physics of the Hungarian Academy of Sciences, Budapest, Hungary)

The finite lifetime of incandescent lamps is generally explained by the uneven evaporation of the tungsten filament, i.e. by the presence of locally overheated parts, the socalled hotspots. Models for calculation the lifetime are based on the assumption that hotspots are present already in the unused filaments. Such hotspot models are in good agreement with a great part of the experimental results, but in certain cases significant disagreement can be observed. In these later cases the location of burnout is not at the position of the originally largest spot as expected by the hotspot models, but smaller spots or even originally spotless parts of the filament turn to be fatal.

It is to be supposed that not only orignal failures like decrease in wire diameter or emissivity differences can be dangerous but hotspots can be formed at preferred sites of the wire during the life of the lamp.

The present paper deals with changes in microstructure the reason of which can be lead back to the powder metallurgical technology and to the doping of tungsten. The changes may result in formation and growth of hotspots at high temperature.

In regular use of lamps one has to count with at least two characteristic changes in the micropore system of doped tungsten wires: first, the bubble growth as a result of mechanical stress and second, the migration of bubbles as a result of different driving forces.

The bubble-grain boundary interaction plays a dominant role in both cases. The mobility of bubbles depends very much on the size, therefore in case of large bubbles the-growth, in case of small bubble-size the migration leads to hotspot formation and contributes to the burnout of the lamps.

Paper 36

CLEANING OF TUNGSTEN FILAMENTS BY ION BOMBARDMENT

J M Walls (Department of Physics, University of Technology, Loughborough, Leicestershire)

Low energy ion bombardment is becoming an important technique for producing clean metal surfaces. However, although the technique is effective in producing surfaces which are chmically clean, the precise structural condition of the surface after such treatment is not well established. Moreover, little attention has been paid to the inert gas atoms which may be implanted into the surface and sub-surface region.

In this paper experiments using the field ion microscope and other surface techniques will be described in which the effects of sputtering tungsten surface with neon, argon and xenon ions in the energy range 100eV to 1 keV have been studied, under conditions similar to those commonly used during sputter-cleaning schedules. In this way the nature and extent of the structural damage caused by each ion species has been compared and data is also presented on the depth of point defect damage caused as a function of ion energy. This information is then used to make recommendations for the most effective means of sputter-cleaning consistent with the amount of lattice disorder produced.

MAGNESIUM ALUMINATE SPINEL AS LAMP ENVELOPE MATERIALS

P Hing (Thorn Lighting Ltd., Melton Road, Leicester)

The fabrication of translucent polycrystalline magnesium aluminate spinel for lamp envelopes by conventional sintering has been undertaken. Suitable spinel powders have been obtained by fusion and calcination of high purity magnesium nitrate hexahydrate and aluminium ammonium sulphate dodecahydrate. It is found that spinel with a slight deviation on both sides of the stoichiometric composition (71.80% Al_2O_3, 28.20% MgO) can be sintered to translucency at about 1700°C in a free flowing hydrogen atmosphere. The composition very rich in magnesia on the stoichiometric side of the spinel, however, is difficult to sinter to translucency. The grain size decreases markedly as the magnesium oxide content increases. Small addition of free magnesium oxide has a beneficial effect on the translucency of the spinel. Possible sintering mechanisms in non-stoichiometric spinel are discussed.

A comparison of the grain-growth kinetics of alumina and spinel shows that the latter follows a square root whereas the former follows a cube root dependence of time. The rate constant for spinel in the region of 1600-1700°C is at least two orders of magnitude higher than alumina for starting materials of comparable purity. The analysis and preliminary experiments suggest that large scale production of tubular spinel within a practical sintering time can be achieved with optimisation of the powder processing parameters.

The predominant optical defects in sintered spinel are white spots identified as clusters of grains smaller than those of the bulk sample with the pores located mainly at the grain-boundaries. The elimination of white spots formation is being investigated.

Investigations on the corrosion resistance of spinel in sodium and other metal halides vapours will be mentioned. Microstructural design for improving the strength as well as the overall in-line transmission of polycrystalline optical ceramic materials will be discussed.

The possibility of using more refractory and corrosion resistant materials of higher translucency than alumina, which is susceptible to sodium vapour attack with the formation of β-alumina, has opened new avenues of designing high pressure sodium lamps. Some of the designs could therefore be successfully exploited. Investigations on the corrosion resistance of spinel in sodium and other metal halides vapour will be mentioned.

Microstructural design for improving the strength as wellas the overall inline transmission of polycrystalline optical ceramic materials, in particular, magnesium aluminate spinel will be discussed.

Paper 38

INFRARED REFLECTING FILTERS AND THEIR APPLICATION IN LIGHT PRODUCTION

H Kostlin, (Philips Forschungslaboratorium Aachen GmbH, Aachen)

Thin In_2O_3-layers have been used for several years in Na-low pressure discharge lamps to decrease the energy consumption. They transmit the typical yellow Na-radiation but reflect the heat radiation back to the discharge tube. In this way they enable the efficiency to be increased to 200 lm/W, which is up till now the higest value of any light source.

The principle of saving energy by selective relection of the unwanted heat radiation back to the source can be applied in many light sources, however under the condition that a suitable filter can be manufactured. This filter can also be used simply to transmit only the visible radiation and thus in the case of lamps with a high amount of heat radiation can be succesfully applied for the production of cold light.

Two types of filter are of importance. Both are based on thin film techniques. On the one hand is the plasma filter, which consists of a thin layer of a highly dopable semiconducting material such as In_2O_3. This is transparent in the visible spectral region, but reflecting in the infrared like a metal. This property is correlated to the plasma oscillation of the free electrons resulting from the dope. On the other hand is the interference filter which consists of a stack of dielectric layers with alternating refractive index. In this case high transparency in the visible region and high reflection in a limited infrared region can be obtained. The possibilities and limitations of these filters are discussed.

By combining a more sophisticated In_2O_3-plasma filter and a suitable interference filter an especially interesting filter can be achieved. Its range of high infrared reflection starts immediately on the border of the eye's long-wavelength sensitivity and extends into the far infrared. In the visible region however good transparency persists. This combination filter is especially efficient for obtaining cold light from a incandescent lamp and by a suitable arrangement can increase the efficiency of the lamp.

Foundation for the Advancement of the Science & Technology of Light Sources
FAST-LS
Belmayne House
99 Clarkehouse Road
Sheffield, United Kingdom
S10 2LN

ISBN 978-1-5108-4106-2

CURRAN ASSOCIATES INC.
proceedings
.com

9 781510 841062

Fifteenth International Conference on Advanced Cognitive Technologies and Applications (COGNITIVE 2023)

Held at ComputationWorld 2023

Nice, France
26-30 June 2023

IARIA